WITHDRAWN
BY
WILLIAMSBURG REGIONAL LIBRARY

SPOTLIGHT ON SPACE SCIENCE

JOURNEY TO PLUTO AND OTHER DWARF PLANETS

KELLIE STEWART

New York

Published in 2015 by The Rosen Publishing Group, Inc.
29 East 21st Street, New York, NY 10010

Copyright © 2015 by The Rosen Publishing Group, Inc.

All rights reserved. No part of this book may be reproduced in any form without permission in writing from the publisher, except by a reviewer.

First Edition

Editor: Susan Meyer
Book Design: Kris Everson

Photo Credits: Cover (main), p. 5 NASA, ESA and G. Bacon (STScI); cover (small image of Pluto), pp. 9, 11, 16, 17, 19, 23, 29 NASA; p. 7 NASA, C.R. O'Dell and S.K. Wong (Rice University); p.13 http://en.wikipedia.org/wiki/Lowell_Observatory#mediaviewer/File:Lowell_Observatory_-_Clark_telescope.jpg; p. 15 ESO/L. Calçada; p. 21 CalTech; p. 25 NASA/JPL-Caltech; p. 26 NASA/Johns Hopkins University Applied Physics Laboratory/Southwest Research Institute; p. 27 NASA Kennedy.

Library of Congress Cataloging-in-Publication Data

Stewart, Kellie.
Journey to Pluto and other dwarf planets / by Kellie Stewart.
p. cm. — (Spotlight on space science)
Includes index.
ISBN 978-1-4994-0375-6 (pbk.)
ISBN 978-1-4994-0404-3 (6-pack)
ISBN 978-1-4994-0426-5 (library binding)
1. Dwarf planets — Juvenile literature. 2. Pluto (Dwarf planet) — Juvenile literature. I. Title.
QB701.S74 2015
523.48—d23

Manufactured in the United States of America

CPSIA Compliance Information: Batch #CW15PK: For Further Information contact Rosen Publishing, New York, New York at 1-800-237-9932

CONTENTS

NO LONGER A PLANET 4
HOW DO DWARF PLANETS FORM? 6
OUR SOLAR SYSTEM 8
PLUTO'S ORBIT 10
PLANET X NO MORE! 12
A LOOK INSIDE 14
IS CHARON A MOON OR PLANET? 16
ERIS, HAUMEA, MAKEMAKE, AND CERES 18
MYSTERIOUS ERIS 20
ASTEROID, PLANET, OR BOTH? 22
A CLOSER LOOK AT CERES 24
NEW HORIZONS 26
IN SEARCH OF NEW DWARF PLANETS 28
GLOSSARY 30
FOR MORE INFORMATION 31
INDEX 32

NO LONGER A PLANET
CHAPTER 1

At the time of its discovery in 1930, Pluto was considered the ninth **planet**, as well as the smallest and most distant known planet in our **solar system**. Since 2006, however, Pluto has been getting a lot of attention for not being a planet after all!

When other objects similar to Pluto in size were discovered in the outer reaches of the solar system, astronomers began to question if little Pluto actually qualified as a planet. In 2006, the International Astronomical Union (IAU), a group of astronomers whose job is to decide how objects in space should be labeled, or classified, made a decision. Pluto would be reclassified as a **dwarf planet**.

Since 2006, there has been a lot of discussion and disagreement among astronomers and space fans about the decision to change Pluto's classification from planet to dwarf planet. What no one disagrees about, however, is that this tiny, distant, icy world remains as fascinating to study and explore as it ever was!

This artist's drawing shows the dwarf planet Pluto and its moon Charon as they might be seen from one of Pluto's smaller moons.

HOW DO DWARF PLANETS FORM?
CHAPTER 2

About 4.5 billion years ago, the Sun, Pluto, Earth, and everything in the solar system did not exist. The chemical ingredients needed to make a solar system did exist, however. These ingredients were floating in a vast cloud of gas and dust called a **nebula**.

Over millions and millions of years, part of the cloud began to collapse on itself. Gas and dust collected, creating a massive sphere, or ball. As the sphere rotated in space, the remaining gas and dust formed a disk that swirled around the middle. Pressure and extreme heat built up as the material in the sphere was pressed together by **gravity**. Finally, the pressure and heat became so great that the sphere ignited to become a new **star**. That star was our Sun!

Leftover matter from the formation of the Sun continued to spin in the disk. Over time, this

The Orion Nebula contains over a hundred planetary discs. These discs could one day form solar systems like our own.

matter clumped together to form planets, moons, **asteroids**, and dwarf planets, including little Pluto. From the moment each of these objects formed, they have been circling, or **orbiting**, the Sun.

OUR SOLAR SYSTEM
CHAPTER 3

Classifying the different objects in the solar system can be confusing. So here's a quick guide to what's what in the solar system. A planet is a large, round object that orbits the Sun. A "true" planet, like Mars or Earth, is so large that it has cleared its orbit, or pathway, around the Sun.

A dwarf planet also orbits the Sun and has a rounded shape. However, because it is much smaller than a true planet, a dwarf planet is not powerful enough to clear its orbital pathway of other space objects. So a dwarf planet will be found orbiting the Sun on a pathway that it shares with many other similar objects.

A moon is a rounded object that is held in orbit by the gravity of a planet, so it forever orbits that planet. Moons can be just a few miles (km) across or as big as a planet. For example,

Jupiter's four largest moons, shown here, are moons because they orbit a planet. Dwarf planets orbit the Sun.

Jupiter's huge moon Ganymede is bigger than the planet Mercury. Ganymede can't be called a planet, though, because it is not freely orbiting the Sun, but is trapped in orbit around Jupiter.

PLUTO'S ORBIT
CHAPTER 4

Pluto is orbiting the Sun in a region of the solar system called the Kuiper Belt. The Kuiper Belt is a region beyond the orbit of Neptune where many icy objects orbit.

Pluto is so far from the Sun that to make one full orbit, it has to travel on a journey of 22.6 billion miles (35.5 billion km). The time it takes a planet to orbit the Sun once is called a year. Earth takes 365 days to make one full orbit, so an Earth year is 365 days. Pluto needs an incredible 90,553 days, however, to make one orbit of the Sun. So a year on Pluto lasts for 248 Earth years!

Pluto's orbit around the Sun is elliptical, or oval shaped. This means Pluto's distance from the Sun changes as it moves through its orbit. At the furthest point in its orbit, Pluto is about 4.5 billion miles (7.2 billion km) from the Sun.

Pluto's elliptical orbit takes the dwarf planet through the Kuiper Belt. However, at other points, Pluto's orbit moves inside that of the planet Neptune.

PLANET X NO MORE!
CHAPTER 5

In the early 1900s, astronomers began searching for a ninth planet beyond Neptune.

Astronomers believed there was another body, which they called "Planet X," whose gravity was having an effect on Uranus. In 1915, this search produced two faint images of Pluto, but the astronomers who took the pictures didn't recognize what they had found.

In 1929, the Lowell Observatory in Arizona instructed a young American astronomer named Clyde Tombaugh to step up the search for Planet X. Tombaugh carried out this search by taking matching sets of photographs of the same area of the night sky two weeks apart. Then he looked at the sets to see if any objects had changed their position.

The Lowell Observatory was established in 1894 in Flagstaff, Arizona.

Finally, in 1930, Tombaugh detected a moving object in one set of images. Soon, the world was introduced to Pluto. For nearly 70 years, Pluto would be considered the ninth planet in the solar system.

A LOOK INSIDE
CHAPTER 6

Because Pluto is so far from Earth, scientists still have much to learn about this dwarf planet.

Scientists believe that Pluto has a solid core made of rock and metal. This core is surrounded by a layer called the mantle that is made up of water and ice. The hard outer crust of Pluto is made of frozen methane and nitrogen.

During times when Pluto's orbit brings it closer to the Sun, ice on the dwarf planet's surface warms and evaporates to become gas. The gases rise from the dwarf planet's surface and form an atmosphere. As Pluto moves further away from the Sun again, though, the gases freeze and fall back onto Pluto's surface as ice.

Pluto is very small, so it only has about 8 percent of the surface gravity that we experience here on Earth. This means if you

Scientists believe Pluto has an icy surface and a thin, hazy atmosphere.

weigh 100 pounds (45 kg) on Earth, you would only weigh 8 pounds (3.6 kg) on Pluto.

IS CHARON A MOON OR PLANET?
CHAPTER 7

In 1978, American astronomer James Christy discovered what for many years was considered Pluto's only moon. It was named Charon after the ferryman in Greek myths who carried the souls of the dead across the river Styx to the kingdom ruled by the god Pluto.

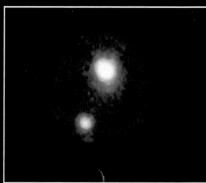

This photo taken by the Hubble Space Telescope shows Pluto and Charon.

Charon is a small moon compared to Earth's moon, but its diameter of about 750 miles (1,200 km) is just over half the size of Pluto's. This makes Charon a large moon compared to its "parent" planet.

When Charon was discovered, Pluto was still classified as a planet. Charon's size was so close to that of Pluto that scientists considered classifying

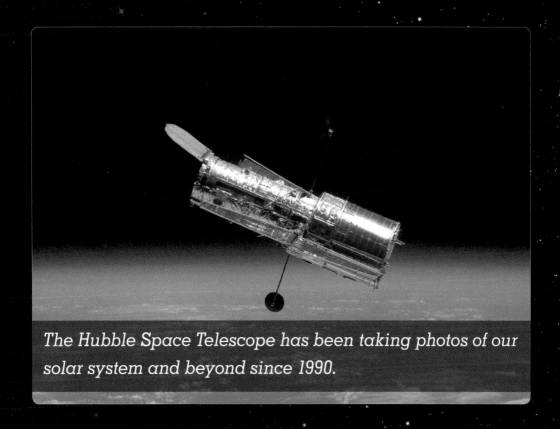

The Hubble Space Telescope has been taking photos of our solar system and beyond since 1990.

the two bodies as a binary, or double, planet. Since Pluto was declared a dwarf planet, scientists have been discussing if the two bodies should now be considered a "dwarf double planet."

Charon is 12,200 miles (19,640 km) from Pluto. Like all moons, it is orbiting its parent body. To make one full orbit of Pluto takes Charon 6.4 days.

ERIS, HAUMEA, MAKEMAKE, AND CERES

CHAPTER 8

As of 2014, five dwarf planets have been recognized by the International Astronomical Union. They are Pluto, Eris, Haumea, Makemake, and Ceres.

Eris, Haumea, and Makemake are orbiting the Sun in the Kuiper Belt, like Pluto. Ceres, which is an asteroid and a dwarf planet, orbits much closer to the Sun, between the orbits of Mars and Jupiter, in a region known as the **asteroid belt**.

Egg-shaped Haumea is about the same size as Pluto. It is one of the fastest-spinning solar system objects yet discovered, making one full rotation on its axis every four hours! Haumea is nearly twice as wide at its **equator** as it is from one pole to the other. Scientists think

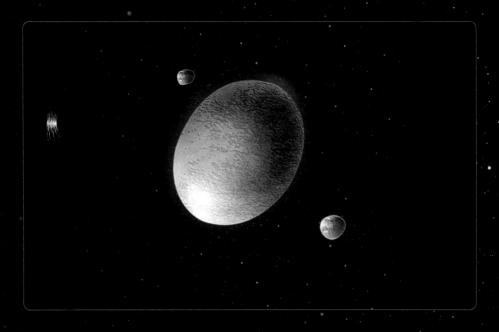

Haumea, seen here with its moons Hi'aka and Namaka, is located outside of Neptune's orbit.

that Haumea spins so rapidly on its axis that material around its equator bulges outward.

Scientists believe Makemake is slightly smaller than Pluto. This dwarf planet takes 310 years to make one full orbit of the Sun.

MYSTERIOUS ERIS
CHAPTER 9

The dwarf planet Eris was discovered in 2005 by a team of astronomers at Palomar Observatory in California.

Eris's orbit takes it about three times farther from the Sun than Pluto and nearly 97 times farther than Earth. Orbiting the Sun with Eris is its moon, Dysnomia.

Eris is so far from the Sun that it takes 557 years for it to make a single orbit. Scientists think that surface temperatures on the icy dwarf planet may only reach −359°F (−217°C). At first, scientists thought that Eris might be larger than Pluto. Now it's believed it may be slightly smaller, but no one knows for sure. There are plenty of mysteries to be solved.

Eris is seen in the foreground with its moon Dysnomia nearby. Far off in the distance is the Sun.

Dysnomia

the Sun

Eris

ASTEROID, PLANET, OR BOTH?
CHAPTER 10

Asteroids orbit the Sun in many parts of the solar system. Today, over 500,000 have been found and studied.

Most of the solar system's asteroids are found in a huge area known as the asteroid belt. The asteroid belt is between the orbits of Mars and Jupiter. In this area, millions of asteroids form a vast, doughnut-shaped ring. The dwarf planet and asteroid Ceres is found in this region.

Ceres was discovered on January 1, 1801, by Italian astronomer Giuseppe Piazzi. When it was first discovered, it was classified as a planet. Some astronomers felt that Ceres was the "missing planet" between Mars and Jupiter, and it was listed as a planet in astronomy textbooks until the 1850s. The discovery of other objects in Ceres's orbit eventually led astronomers to understand, however, that Ceres wasn't a true planet. It was reclassified as an asteroid, and today, it also belongs to the dwarf planet club.

While the asteroid belt contains hundreds of thousands of asteroids, most are very small. The asteroid Ceres alone makes up more than a quarter of the mass of the entire asteroid belt.

A CLOSER LOOK AT CERES
CHAPTER 11

When an asteroid is discovered and its orbit has been studied and recorded, it is given a number and sometimes a name. Ceres, the first asteroid to be discovered, is officially named 1 Ceres.

Ceres was the first asteroid to be found because it was the easiest to see from Earth! Ceres has a diameter of about 600 miles (966 km), which is almost as far across as the width of Texas. It is by far the largest object in its orbital pathway, and its mass makes up about one-quarter of the total mass of the asteroid belt.

Ceres has a central core of hard, rocky material, a mantle, and a rocky outer crust. Scientists believe the mantle, which is 62 miles (100 km) thick, is made up of ice. If this is correct, it would mean that Ceres contains more water than all the freshwater on Earth! Just like Earth, Ceres may also have ice at its north and south poles.

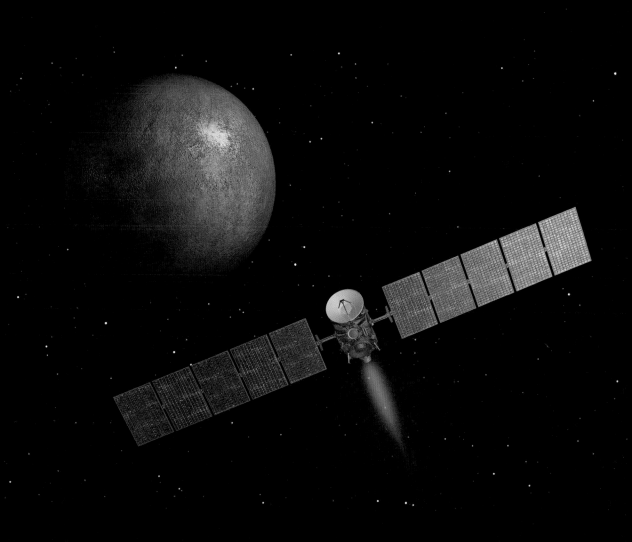

The Dawn spacecraft was launched in 2007 to study the asteroid Vesta and the dwarf planet Ceres.

NEW HORIZONS
CHAPTER 12

On January 19, 2006, the robotic spacecraft *New Horizons* roared into space atop an Atlas V rocket from Cape Canaveral, Florida.

This is the image of faraway Pluto captured by *New Horizons* in 2006. The arrow indicates the position of Pluto.

If the mission is successful, *New Horizons* will be the first space probe to perform a flyby of Pluto and its five known moons. By 2014, the craft had already flown past the orbits of Mars, Jupiter, Saturn, Uranus, and Neptune. It was expected to reach Pluto by July 14, 2015.

Although *New Horizons* was still in the early stages of its journey, it created its first images of Pluto in September 2006. These pictures, taken

New Horizons *began its journey to Pluto from Cape Canaveral on January 19, 2006.*

from a distance of about 2.6 billion miles (4.2 billion km) from Pluto, showed that the probe could track objects across huge distances.

IN SEARCH OF NEW DWARF PLANETS
CHAPTER 13

New Horizons should fly within 6,200 miles (10,000 km) of Pluto and 17,000 miles (27,000 km) of Charon. At those close-up distances, the images of our most famous dwarf planet should be extraordinary.

As of 2014, plans hadn't been finalized for *New Horizons*. However the goal was to send the probe deeper into the outer solar system following its Pluto flyby. Starting in 2016, *New Horizons* will begin flybys of other objects in the Kuiper Belt, which is the home of Pluto and what may be hundreds of thousands of other objects. As of 2014, the mission is scheduled to run until 2026.

The objects that *New Horizons* will study in the Kuiper Belt may include many that will eventually become classified as dwarf planets. Therefore, the *New Horizons* mission may change the way we view the make-up of our solar system for years to come.

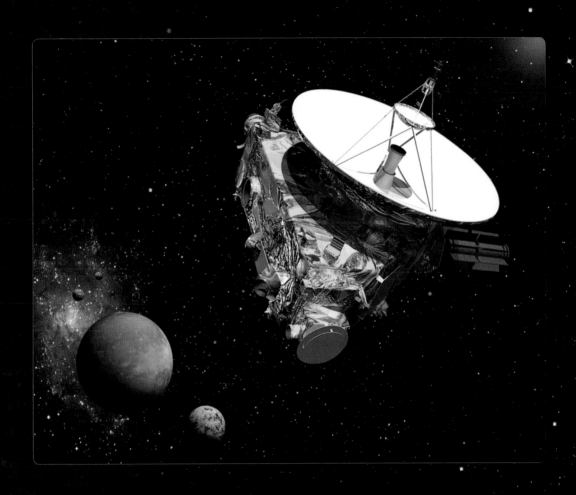

This artist's drawing shows what it might look like when New Horizons approaches Pluto and its moons.

GLOSSARY

asteroid: A small, rocky, planet-like body in space that circles the Sun.

asteroid belt: The region of space between the orbits of Mars and Jupiter in which most asteroids are found.

dwarf planet: A body in space that orbits the Sun and is shaped like a sphere but is not large enough to disturb other objects from its orbit.

equator: An imaginary circle that divides a planet into two equal parts.

gravity: The attraction of the mass of a body in space for other bodies nearby.

nebula: A huge cloud of dust and gas found between stars.

orbit: To travel in a circle or oval around something. Also, the path used to make the trip.

planet: A large body in space that has its own motion around the Sun or another star.

solar system: The Sun and the space bodies that move around it, including the planets and their moons.

star: Any one of the objects in space that are made of burning gas and that look like points of light in the night sky.

FOR MORE INFORMATION

BOOKS

Portman, Michael. *Why Isn't Pluto a Planet?* New York, NY: Gareth Stevens Publishing, 2013.

Roza, Greg. *Pluto: The Dwarf Planet.* New York, NY: Gareth Stevens Publishing, 2011.

Simon, Seymour. *Our Solar System.* New York, NY: HarperCollins Publishing, 2014.

WEBSITES

Due to the changing nature of Internet links, PowerKids Press has developed an online list of websites related to the subject of this book. This site is updated regularly. Please use this link to access the list: www.powerkidslinks.com/soss/pluto

INDEX

A
asteroid belt, 18, 22, 23, 24
asteroids, 7, 18, 22, 23, 24, 25
atmosphere, 14, 15

C
Ceres, 18, 22, 23, 24, 25
Charon, 5, 16, 17, 28
Christy, James, 16
core, 14, 24
crust, 14, 24

D
Dawn, 25
dust, 6

E
Eris, 18, 20, 21

G
gas, 6, 14
gravity, 6, 8, 12, 14

H
Haumea, 18, 19
Hubble Space Telescope, 16, 17

I
International Astronomical Union, 4, 18

K
Kuiper Belt, 10, 11, 18, 28

L
Lowell Observatory, 12, 13

M
Makemake, 18, 19
mantle, 14, 24
moons, 5, 7, 8, 9, 16, 17, 19, 20, 21, 26, 29

N
nebula, 6, 7
New Horizons, 26, 27, 28, 29

P
Palomar Observatory, 20
Piazzi, Giuseppe, 22
Planet X, 12

S
solar system, 4, 6, 7, 8, 10, 13, 17, 18, 22, 28

T
Tombaugh, Clyde, 12, 13